BATTLE OF THE STARS AND THE FIERY IMPACT OF PHAETHON SEEN AS NATURAL HISTORY

BATTLE OF THE STARS AND THE FIERY IMPACT OF PHAETHON SEEN AS NATURAL HISTORY

FRANZ KUGLER
Prism Thomas

G. Stempien Publishing Company

Battle of the Stars and the Fiery Impact of Phaethon Seen as Natural History
By
Franz Xaver Kugler

Translated and expanded by
Prism S. Thomas

ISBN 978-0-930472-65-8
G. Stempien Publishing Company
Copyright © 2022 by Prism S. Thomas
All right reserved

This book is a translation and
an interpretation of an earlier work
by Father Franz Xaver Kugler
(Munster 1927).
The intention is to infuse
this scientific study with the
imagination and force which
pure mathematics and astronomical
formulae could not alone convey.
The author
Winter Haven (USA), Easter 2022

CONTENTS

Translator's Introduction

Kugler's Introduction: The Stars as Fighters

1. THE SUPPOSED "MAD FINALE" OF THE 5TH BOOK OF THE SIBYLLINE ORACLES: A MEANINGFUL NATURAL POETRY

2. EXPLANATION OF THE TEXT: THE PRELUDE TO THE STAR WAR

3. WHO OR WHAT IS PHAETHON?

4. THE STAR STRUGGLE OF DIONYSIACA AND THAT OF SIBYLLINA.

5. PHAETHON'S FALL INTO ERIDANUS (CELESTIAL RIVER) WORLD FIRE AND DEUCALION (GREEK NOAH) FLOOD.

TRANSLATOR'S SYNOPSIS

TRANSLATOR'S INTRODUCTION

This book is about a true astronomical catastrophe that occurred between 1500 B.C. and 1100 B.C. which engulfed part of the planet under water and scorched other portions of the world in flame. Although the event was real, its occurrence had been disguised by mythology into a fable. But, the Babylonians and others documented the heavenly changes when this cataclysm occurred and left a record in their ancient cuneiform texts. They did this through means of astrological illusions and severely complex astronomical calculations so that it required a genius in both areas to interpret them.

This genius was the Jesuit Father Franz Xaver Kugler. He spent his life translating thousands upon thousands of cuneiform tablets and then he performed the mathematical calculations that verified the locations of the astronomical events and their effects on the planet earth. He finally published his findings in 1927 and it is this short book that is translated here. Because the book lacked an understandable "storyline" the translator has taken that task upon himself and produced a cosmic tale (though true) that follows closely the outline of the astronomical facts.

This true "cosmic tale" should not be confused with the many other fictional stories that have been erected surrounding the cosmic catastrophe that struck the earth in its distant, though historic, past. Many of these stories involve the destruction of the speculative planet that existed between Mars and Jupiter. The ancients were unaware of the asteroid zone between Mars and Jupiter and the theory that it was populated by the fragments of an exploded planet was first introduced by the German astronomer, Heinrich Olbers (Heinrich Wilhelm Olbers (1758–1840).

After Olber's theory became common knowledge, many of the Phaethon stories involved the destruction of the former planet that orbited between Mars and Jupiter. In one version the planet was destroyed by nuclear war. In another it was shattered by colliding with an asteroid or solar like fragment that intruded into the solar system. And another theory states that God destroyed this evil planet between Mars and Jupiter and that the survivors - who were devils - made their way to earth or mars and colonized one or both of these planets. None of these stories are verified by factual evidence. The original Phaethon story did not account for the area of origination of the intruding objects.

Some of the wording in the original Kugler document is archaic, of a poetic nature, and not easily deciphered. The first part of this work is heavily astrological and poetical in style, potent with allegory. It can be enjoyed for its beauty and the background it supplies. The second half illuminates the first with more clarity. A brilliant story is told here; non-fiction, based on the OBSERVED facts in the distant past.

Any missions or errors are my own.

KUGLER'S INTRODUCTION: THE STARS AS FIGHTERS

The word 'army' (sabä) occurs in the Old Testament for the first time in Gen. 2, 1: "Thus was made perfect the heaven and the earth with all their army." Of course it means here not the full totality of the hosts of individual creatures. As a heavenly host in the true sense, the stars only appear much later in the Old Testament and initially as an object of idolatrous worship. The home of this cult was Canaan, the final goal of the Israelite conquest.

The warning against the cult of "sun, moon and the stars, the whole host of heaven" (Deut. 4, 19 and 17, 3) was therefore well founded. In the summary accusation against the kings and the people of Israel (II Kings 17, 15 ff.). It says that they, despite all warnings, imitated the people around them, "made cast images, golden calves, and incense trays, worshiped the whole host of heaven, and served Baal". Kings 2 ff.). They even erected "altars for the whole host of heaven" in the two forecourts of the Yahweh temple. This cult of the star army certainly also existed before and during the time of the Judges 2:11 ff. and knew of "Baal," a god of the Akkadian and Ugaritic peoples (Trans - Northern Syria) who was closely tied to storms and rain; and "Astarte," who was one of the three great goddesses of the Canaanite pantheon. These are local special forms of the two highest gods.

In addition, Judges 5:20 shows that the stars are seen as warriors. "Heaven's army" is therefore not just an analog, but means a real army of fighters. Where does this idea come from? (Trans - bold in

ALL cases). The orderly movement of the fixed stars and their ranks, which are revealed through the difference in shine and color, already suggest the representation of an army. The apparently random special movement of the planets, which now move forward (to the east), now backward (to the west), now faster, now slower, now in a straight line, then in a tortuous path under the stars, and even stand still for a long time are like scrutinizing commanders.

The change in shape is also extremely striking. The sun and moon, descending from the heavens, reach a significant size and present on the horizon the sight of a stretched out figure like that of a wanderer tired from a long march. The large constellations show a similar change only more strikingly. Against the zenith their limbs appear closer together and on the horizon, on the other hand, enormously extended. They are like such an animal. **"From heaven the stars fought, from their orbits they fought with Sisera."** (Trans - Sisera was commander of the Canaanite army of King Jabin of Hazor).

There is evidently a subtle irony in this. The kings of Canaan had trusted in the army of stars as allies, but the troops of their general met their Death in the waves of Kison, the "torrent stream" along the banks of which the great battle was fought between Israel, led by Deborah and Barak, and the army of Sisera, which — thanks to heavenly decree — had caused a violent thunderstorm and the tides to swell. The Epic of Gilgamesh (XI, 114 ff) probably also reminds us of this: "the gods became afraid of the deluge. They escaped and ascended to the sky of Anu, the lord of constellations and king of gods. Like a dog the gods crouched . . ." Particularly striking is the magnificent Orion, **In Greek mythology,a gigantic, supernaturally powerful hunter who in any case awakens the idea of a mighty war hero and of sea creatures of great freedom of movement and strength who represent the Deluge.**

Even more surprising (especially towards morning) are the shooting stars that traverse the universe like glowing fiery arrows and at certain times of the year make certain constellations the starting point of a sparkling rain of fire. And when in the cloudless sky great flares, sometimes with a hissing noise, sometimes with thunder and lightning - sometimes causing earthly fires - are visually and audibly experienced at the same time.

It is also understandable that a naive view of nature interprets the eclipses of the moon and sun as oppression by a power hostile to the light. The retreat of the darkness is the final victory of the great divine heavenly lights over the demons. The rarer, grander, and more enduring the phenomenon, the more lasting the effect on the mind. Great comets, therefore, by their unexpected appearance and rapid growth, their gigantic extent and long duration, have always been a celestial sign of impending frightful events. And all this abundance of power revealed in the starry sky is increased by the connection with the shattering forces of nature in the clouds. In the case of the ancients, who did not know the enormous distance between the stars and the earth, the appearances of both kingdoms flow into one another. This is particularly so in the Babylonian heavenly vision. Thus the weather god Adad and the sun god Samas appear like two heroes side by side. "Well, let us build for ourselves a city and a tower, the top of which reaches to heaven..."

On one and the same stage, their roles complement each other not only in the organic life of nature, but also in the religion of the stars and the interpretation of the cosmos. It was memorable to have strange phenomena happen around the time of the full moon like the halo rings along with parhelia (below)

PARHELIA

These sights evoked warlike ideas, especially when the planet (Jupiter) was in the middle of the ring and the representative of a hostile power (Mars) approached him from outside: a picture of the siege and the threat of invasion.

These examples may suffice. They explain the view of the ancients that the starry sky is not only alive, but also a scene of warlike complications. And the following pages also relate to this. Their subject is about the battle of the stars, about which the Sibylline oracles in Book 512 and 206 report on the Alexandrian Phaethon saga, which the humanistically educated person already read in Ovid's Metamorphoses (43 B.C. - 17 AD).

For us, however, we are not concerned with the appreciation of the poetic clothing, but with the actual, natural-historical basis of the two ancient written classics. Up to now an impenetrable mist has hovered over the finale of Book V of the Sibylline Oracles (Vs 512 ff.), and the secret of Phaethon was kept mostly hidden.

But once the first and most difficult riddle has been solved, we have also found the key to the second. Both will turn out to be fact in the course of our investigation. Whether our writing can also be considered "contemporary" is questionable but It is certainly of cultural and historical significance. Above all, it imparts the powerful lesson that ancient traditions, even in the guise of myth or legend, should not be easily dismissed as fantastic or even as meaningless creations. And this caution is all the more appropriate where serious reports of a particularly religious nature are in question, such as are found in abundance in the Old Testament in particular. It is precisely here that the resolution for strict self-criticism would be extremely helpful and highly up-to-date: first weigh, then dare! We will soon substantiate the urgency of this warning with numerous documents in this special publication.

Valkenburg (Holland), Easter 1927. The author.

1. THE SUPPOSED "MAD FINALE" OF THE 5TH BOOK OF THE SIBYLLINE ORACLES: A MEANINGFUL NATURAL POETRY

The end of the fifth book (Vs 512-531) of the famous Sibylline oracles forms the prophecy of a "war of the stars" which ends with the fact that they will be hurled down and will set the whole earth on fire. Blass calls the passage a "mad finale". The great contempt with which Blass also otherwise speaks of his shock makes this devastating judgment appear quite suspicious. It differs little, however, from what the Oracle's' best known recent arranger—Job Geffcken—said. "The tumbling imagination of this vision, although the poet operates with astrological terms, mocks every astrological system. A mathematician, E. Hoppe, to whom I showed the passage, also explained to me (Kugler) that, viewed from every point of view, it contained complete nonsense." From this one may well conclude that until then no one had succeeded in recognizing any meaningful allegory in the war of the stars, let alone real cosmic events.

However, many years of occupation with the deciphering of cuneiform inscriptions about the astrological and astral-mythological views of the Babylonians has taught me that very much of what appears nonsensical to us modern Westerners about the oriental and especially the ancient oriental world of ideas is not lacking in healthy logic.

CUNEIFORM (In use from 3200 B.C. to 200 C.E.)

This knowledge was also the reason to at least attempt a solution to the pending question. The success surpassed the boldest expectation as the "mad finale" revealed itself as a pretty disguise of real natural events according to a perfectly unified plan. And so faithfully does the poet reproduce the astronomical processes that the modern tools of calculation not only mark the homeland of the poetry's origin, but even determine the season to which it corresponds. The celestial drama begins with an interesting prelude, which at the same time establishes the actual star battle. Two mighty meteors of apparent size and shape of the sun and the moon (as they are seen at a distance) appear threateningly in the sky with their characteristic accompanying phenomena in turmoil; and the actual star battle begins.

FLAMING OBJECT

Below - (The morning star, Venus, standing on the back of the lion, (astrological Leo Constellation), initiates the battle).

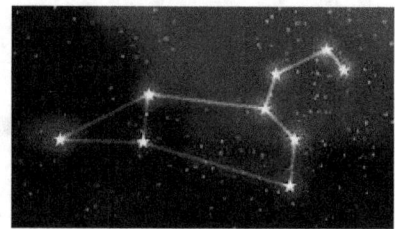

LEO (Venus would be above its back)

The battle causes a complete upheaval from Virgo (constellation) to the middle of Aries (constellation). The stars that ruled the dawning sky at the beginning of the battle finally descend into the Oceanus (the river flowing around the world), setting the earth on fire.

This is the basic idea in poetical fashion, faithfully recorded in every detail, but recognized only after the facts are laid bare, which the veil of poetry hides from the profane gaze. In undertaking this prosaic work, we cannot, above all, do without the astronomical calculation. It alone can give us certainty about the astronomical positioning. What is certain is that Venus stood as the morning star on the "lion's back" when the battle began; the sun was therefore east of this in Virgo, which is also attested by a parallel written source. But if we assume that "the sun at that time happened to be in the middle (on the 15th) of the sign of Virgo" this estimate is justified, according to which, at the beginning of the battle the moon was in its old light (= last crescent just before the new moon) showed.

But at the end of the battle, the interrelationship between the sun and the fixed stars, would give a time period of two thousand years ago. But all of this information is necessary in order to recognize the factual meaning of individual passages and only then are we in the position to decide certain questions critical of the text and to offer an accurate translation, allowing us to understand the acquaintance with the meteoric phenomena as well as the Greek and Oriental astrology that is conveyed by short explanations and corresponding documents.

2. EXPLANATION OF THE TEXT: THE PRELUDE TO THE STAR WAR

(Trans - This entire section is devoted mostly to astrological terminology and can be challenging for a person who isn't an expert in astrology. However, the primary purpose is to set a date and a place for the actual occurrence of the inferno and deluge that is the topic of this work and in that matter this section succeeds).

A shining 'sun' menaced the stars and I saw a 'moon' of terrible wrath flashing with lightning bolts. The stars were weaponized. God made them fight. Instead of the normal 'sun' there appeared only long flames tangled up. It is certainly not the day star here; for it appears in a position among the stars. It is in truth a sun-sized meteor, which menacingly lights up the sky. Such a meteor of the apparent diameter of the sun or the moon has often been observed. The Babylonians long ago called it a Samsu 'sun'. Its shape, however, is by no means always circular or spherical, but quite variable and often resembles the partially illuminated lunar disk.

And such a thing was certainly spoken of before. This is because of the connection with a previous great meteor, and even more so because of the "terrible wrath" of lightning which followed. The actual moon is a peaceful wanderer which does not cast lightning or roar but **The meteor "moon" is different; it can do both. Sometimes one sees flashes of lightning, but usually one also hears the mighty thunder**

of the bursting heavenly explosions. Understandably, this also causes the stars to tremble and extends the destruction. Instead of the (meteor) sun, the seer perceives "long flames" passing through and entwining with each other. What does that mean? The meteor has disappeared or has exploded into pieces in the atmosphere of which there are numerous verified cases in history as described by Arago in his publication of 1860 in "Popular Astronomy" on page 230.

The Babylonians paid attention to this early on; which the reader can see in my submissions to Astronomy Magazine and in my book "Astronomy in Babel" pp. 89-92. These explosions were so ordinary that sometimes they repeated themselves and even developed into formal cannonades (see, for example, the statements by Arago from the years 1643 and 1651). This is how the repeatedly observed showers of stones come about, the embers of which have caused large fires as they crashed glowingly through each other, leaving long, luminous and crisscrossing tails.

THE ACTUAL STAR FIGHTING.

"The morning star directed the battle by mounting the lion's back." (The Lion is the constellation of Leo: its back is the eastern portion of the constellation). The equation that was given translated into the position of Venus as the morning star and this calculation has been assumed to be accurate. The afterglow of meteors is something quite ordinary. How attentively Seneca heeded such and other unexpected celestial events is shown, among other things, by nature where, remembering their powerful impression on the human heart, he emphasizes the lasting effects produced. According to the Egyptians and Homer, Plato and Achilles Tatius, the star or great comet is identified as Venus and as the morning star.

The Babylonians had recognized this at the time of King Ammizaduga (1801-1780 BC - 4th ruler after Hammurabi), as can be clearly seen from the Venus texts I have edited in my book "Babel II" 257 ff. However, this much also emerges from the quoted passage that the identity of the morning star in our case requires proof. It can also be brought about like this: Venus stood on the back side (in the eastern part) of the lion. If it was the morning star, the sun had its place in Virgo; otherwise in the western part of Leo or in Gemini. Now, however, from the shorter parallel passage, which we will deal with below, it emerges that the sun was then in Virgo.

The following indirect proof is even more effective. Assuming Venus = morning star, there is perfect harmony between all the following statements characterizing the end of the fight; but the opposite assumption would present us with an unparalleled confusion. **Why is the morning star the leader of the battle and why does he mount the lion as such? Here we encounter ideas whose original homeland is to be sought in Babylon.** Venus as the morning star represented the Istar-kakkabe, "Istar of the Stars" and is also intended as zikarat "male" — in contrast to the Venus evening star, the goddess of love and motherhood.

The morning star thus represents the scientific justification of this mythological double role of Venus as the powerful ruler of the star army in ancient texts. This idea is obviously related to the ancient conception of Venus as the goddess of war (as already at the time of the First Bahel Dynasty of India). But the animal of Igtar-Venus (probably the goddess of battle) is the lion, at the same time the symbol of her power to overthrow everything. This is shown by her representation on the rock relief of Malta, where she stands on a lion and takes the last place. Astrology reveals the intimate relationship of Venus-Istar to Leo; its constellation is her kakkar nissirti (area of her manifest treasure or treasury). The poet, however, had no need to borrow from Babylonian mythology the idea of a battle ready goddess on the back of a lion; for Istar had long since completed her victorious march over the world and erected her throne not only in Syria and Asia Minor, but also in Greece,

Rome and Carthage - albeit under different names and in a somewhat different form.

The Phrygian Cybele, Gaia the Greek mother of the gods, and Tanith of the Carthaginians resemble Venus in many ways; to all, the lion is an inseparable companion. He is the assistant to the throne of Cybele and on his back rides the lance-armed Caelestis. "Of the Song of Songs and its Warlike Bride" "The moon's two-horned mourning shape changed her." However, both changes are initially not required due to the meaning of the passage and their context. Of course one should not translate it "and the moon's two-horned misery exchanged" (Blass) and even less eliminate the resulting nonsense with "and (the morning star!) exchanged itself with the two-horned mourning shape of the moon" (also Blass).

But how is the shape of the moon to be interpreted? It signifies, in fact, the sorrowful, sad state of the Moon, its increasing eclipse before the Moon's conjunction with the Sun. Shortly before it one sees over the eastern horizon the dark disc, dimly lit by the earth's light, half enclosed by the crescent and the horns, the two-horned, mourning figure of the fading moon. This conception was not invented by us, but is of ancient oriental, Babylonian origin.

On the day when the crescent was last visible, the Babylonians held a celebration for Nergal (the god of the dead), and on the following day they held a funeral service for him. About three days after the moon disappears in the east, the young crescent appears in the western evening sky. The happy new light festival of the moon god (Sin) was dedicated to this event. And this change is meant when the poet says: the two-horned Janmier (shape) 'changed'. The crescent moon, which was in the eastern morning sky at the beginning of the star battle, has its place in the western evening sky at the end of the battle, as a symbol of the beginning month and at the same time—as will be shown immediately —of the year.

As already pointed out, the beginning of the struggle coincided with the time when the Sun was casually in the 15th degree of Virgo and the end when it had reached the 15th degree of Aries. **These two dates were (100 BC) 209.4 days or 7 synodic months and 2.7 days apart.** So if the moon was initially in the eastern sky in the disappearance (old light), at the end of the fight it stands as a new light in the western sky. It was then the beginning of a new month for all peoples whose calendar was based on the luna-solar year, and for all those who also — like the Babylonians and Hebrews of the Seleucid and Parthian times — celebrated the New Year in spring. It was also the beginning of the year. Or is it coincidence, or the result of careful choice, that the revolution in the heavens takes place just after seven synodic cycles of the moon? The latter seems more likely to me. "Seven" is the expression of perfection not only in Babylonian but also among the Jews.

It is also worth noting that the moon finally enters that sign of the ecliptic which is known in Greek astrology as the place of its greatest display of power which occurs in Taurus. **The moon is the regular companion of the earth and has nothing to do with the moonlike meteor of destruction.** One also expects the Sibylline Stemkampl (prophecies written in Greek hexameter) and Phaethon undergoing its change at the point where the text offers it, immediately following the appearance of the morning star and the beginning of the battle, because the moon is the main star.

Our explanation adds the further reasons: **the whole turning point in the stellar world begins and ends with the characteristic phases of the moon, the natural determinants of time, and the waning crescent moon is on top of and close to the morning star, which begins the battle.** The ibex (horned goat) then pushed back the young bull's neck; but the bull stole the day of Capricorn's return home. The first scene belongs to the time when the sun is in the middle of Virgo (beginning of the fight). Soon after dark Taurus the bull rises (namely his neck first; the head is turned to the east and lowered); It is this circumstance which gives the poet the idea that the bull has received an enemy blow from the west.

Note the culprit; it is the ibex (the constellation of the horned goat). This is already over the meridian fled west. But he does not escape vengeance. (Trans - the horned goat is known as a sacrificial (scapegoat?)). In the months that follow, as the sun crosses the signs of Libra, Scorpio and Sagittarius, Taurus appears higher and higher at nightfall, following retreating Capricorn until the latter Ibex disappears in the rays of the sun that has just set and lapsed time remains invisible. He is now robbed of the day's return. Here it is already evident that the change in the phase of the moon is not referring to the transition from the old light to the new light that immediately follows, but that both are separated by a period of several months. How many months it is, teaches the following statements in unison and with increasing clarity, all of which refer to the last day of the struggle.

And the scales (Libra) ousted Orion, so that they remained no longer. The explanation of the passage is not easy. Greek mythology, however, connected the scorpion following the scales with Orion, assigning the former to the latter which caused the deadly sting (cf. e.g. Arati Phaenomena 637-646 poetic work of astronomy) - an idea which is known to be based on the fact that Orion sets when Scorpio rises. But Orion and Libra? Is it about the heliacal setting (at sunset) of Libra? The mistaken translation in Blass: "Orion stole the scales" could, however, suggest the assumption. The latter is inadmissible, if only because the poet would certainly not first mention Capricorn setting at the same time as the sun, three months before that of Libra. Moreover, there is no connection between the setting at the same time as the sun of Libra and Orion. No, the matter is quite different. First, we note that at the time (about 100 B.C.) the sun was in the middle of Virgo, the stars of the chariot were the same period after dark as seven months later, when the sun was in the middle of Aries, as also by far most of the stars of Orion.

On the last day of the battle, Orion was on the western horizon at nightfall in almost the same position in the sky as Libra was at the beginning of the battle. This is the meaning of the words: "Orion supplanted the balance so that the weight was distributed." Or, "Orion pushed the scales aside so that they no longer remained". The virgin traded the lots of the Twins in Aries (shifted their positions)." The meaning of this passage is: The role that Virgo played at the beginning of the fight - the sun in Virgo and Venus shining above the sun in the early morning - now is taken by Aries at the end of the fight, after seven months have elapsed. Virgo and Aries have the character of Castor and Pollux, the twins, when Aries rises, and the Virgin disappears at astronomical twilight. Until then, even the 6th magnitude stars were still visible. But ß and y Draconis were already in the last part of the night sky being a full hour below the horizon. It was different at the time when the sun was in Aries. The sun set and astronomical twilight was over; but ß and y Draconis had already risen above the horizon, only to return to it 6 hours 50 minutes after sunrise. Draconis therefore remained above the horizon all night.

The interpretation of the astronomical values leads essentially to the same result; but the contrast between the proportions of the two seasons is much less sharp; for at the time when the sun was in the 15th degree of Virgo, only 19 minutes elapsed from the setting of the star y Draconis to the beginning of the astronomical dawn. And ß Draconis was even only minutes of night below the horizon! However, we have not yet taken into account the effect of atmospheric refraction, which makes sunrise earlier and delays the setting of the stars. If one takes them into account, one can at most speak of the setting of only the star y Draconis, and that only a few minutes before the beginning of astronomical twilight. But this is too inconspicuous to derive a contrast between the two seasons.

"The Pleiades no longer shone." **This statement is very valuable as it gives another means of pinpointing the point at which the battle took place. Now the setting of the Pleiades in geographical latitude 30 took place on April 7th, 100 BC. on April 8 Julian, at a solar longitude of 13 (or 15) degrees.** Apparently the fight was over soon after; otherwise the heavenly poet would have said: the bull or the head of the bull (with the striking star Aldebaran) no longer appeared. **We will therefore have to assume that the last day was around April 8th,** or the day after the setting of r Tauri in the Pleiades at the same time as sunset.

The site of this location is therefore in Egypt (Cairo at latitude 30° and Alexandria about 31°). "The fish crept away opposite the belt of the lion." We have here again a scene from the end times. It belongs to the last part of the night, the beginning of dawn. On the eastern horizon are the recently emerged Pisces, and opposite them on the western horizon is Leo. Owing to the nearness of the Sun, which is in the middle of Aries, the stars of Pisces gradually fade away, first the eastern ones, then the western ones; they "crawl" in the face of the lion (Leo) is the part of the ecliptic belt belonging to Orion. (In any case, this explanation is much more natural than the assumption of a horizontal belt about 30° wide, in which lion and fish face each other and in which the latter fade away.) "The Crab did not stand, it feared Orion." The meaning of the passage is not as follows: Cancer flees from Orion or at sunrise goes inside while Orion becomes visible. Not the former, because Cancer is east of Orion and follows it in daily movement; not the latter, because the crab does not die with sunrise during the entire struggle. Rather, the poet wants to say: **Even Cancer, the embodiment of stability, is swept away by the general upheaval; for better or for worse, he follows the bidding of the mighty warrior Orion, who is not far from him.**

Several philologists have labored in vain with this passage. Palely translated: "The scorpion 'attached' the tail to the terrible lion". Lanchester suspects: 'Scorpio drew up his tail because of savage Leo. Geffcken says: "But the tail of the scorpion must stay here"; he reads (the latter with Alexandre) and translates doubtfully: "He (the scorpion) crept behind his tail?" This, however, does not achieve much.

The real sense - and this is what matters above all - can only be found in astronomy. We ask: what temporal-spatial relationship exists between Scorpio and Leo and does this fit into the already recognized plan of the myth? By calculation the following can be determined: After the head and breast of the lion have passed the meridian and the stars of the abdomen culminate in succession, the Scorpio stars d, ß, n through X rise in succession. **The appearance of the head stars d, ß, n occurs in (101 BC) and the geographical latitude 30°.**

Any deviations from the real time and the real place of observation are irrelevant for the result. Consequently, the rising of Scorpio's Head is the first clear star appearance on the eastern horizon, while right in the middle of the sky stands the terrible Leo. Further, just where the last (tail) star ß of the latter passes through the meridian, the last (tail) star X of Scorpio rises on the eastern horizon. This is the first act of the nocturnal spectacle. The second is this: The descending lion is followed by the scorpion, whose head and chest sink while the scorpion's tail reaches the highest point on the meridian. Thus the complete setting of Leo is the last noticeable event on the western horizon. The whole night is a witness, again scorpion chasing the lion or its tail. This explains the real meaning of our passage completely.

However, it can be corrected linguistically without obligation; this is done by combining day and savage hero into one word. "The scorpion goes for the tail of the most terrible lion." The gliding down against the sun is poetically represented as the effect of the blinding or scorching evening glow of the sun. (The possible view that the Sirius star is from the flame of the sun has slipped is hardly permissible, since when it appears the second twilight arc, which may still be considered, has already sunk x meters below the height of the star and thus could not be regarded as the sliding surface of the latter.)

" **What does Sirius mean here? Alexandre translates it as 'Lucifer' and he is followed by Lanchester and Blass, who translates it into 'Morgenstern'. Nowhere, however, could I find any trace of a justification for this interpretation.** Anders Geffcken calls it Saturn. Of course, this equation has a lot going for it, but it cannot be taken for granted without further ado; for **(The heliacal sinking of Sirius took place at locations of latitude 30 °in 100 BC on May 11; the end of the star battle, however, fell - as shown earlier - on "May 8 or April 7.)** Shining' (usually said of the moon, stars, dawn) is not exactly a property of Saturn, whose Egyptian epithet is herald, (i.e. of destinies in the astrological sense).

Also, "Aquarius cannot burn Saturn." but it does the reverse. Aquarius burned by the faint glow of Saturn, offers a more reasonable idea. If the star in question is really Saturn, the words can hardly be translated and interpreted otherwise than as such: "Aquarius kindled the power of mighty Saturn", ie he gave the planet new luminosity or - speaking quite soberly astronomically - Saturn rose at sunrise in Aquarius and became more and more prominent there during the next month as the morning dawn.

This view would at least not contradict the plan of the whole. The heliacal rising of Saturn in Aquarius could take place 1 to 1/4 months before the end of the battle (position of the sun in the middle of Aries). However, the following positive astrological reasons can also be provided.

1. The epithet points to Saturn, the most powerful of the planets. Accordingly, the particularly frequent use of the name in Pseudo-Manetho.

2. Saturn and Aquarius belong astrologically together; for the latter is the nocturnal (lunar) Home of Saturn

3. **The prophecy of the star war with the following fiery catastrophe refers to Ethiopia.** Now, according to Ptolemy's geography, Aquarius corresponds to central Ethiopia, which is at the same time is under the influence of Saturn (cf. Bouché-Leclercq op.cit. 342 f.) It is therefore to be expected from the outset that Saturn will also appear in conjunction with Aquarius in the constellation, not as an opponent but as an ally .

"Uranus rose up himself until he shook the combatants,hurling them in rage to earth.The Ocean's suddenly erupted but the whole countryside was set ablaze.The ether remained starless."

The situation is as follows. At the beginning of the battle, when the sun was in the middle of the sign of Virgo, at dawn, the constellations Aries, Taurus, Gemini (with Orion), Cancer and Leo were above the horizon. Seven months later, however, when the sun was in the middle of Aries, all these stars had already descended towards morning. At the same time, the hot season also begins in Ethiopia, which brings with it a torrid, scorching heat during the day.

According to the popular-poetic view, however, this is related to the setting of the stars mentioned, for they form the hot region of the ecliptic, unlike the others, which represent the winter and rainy seasons: their descent to the ocean which surrounds the earth sets it on fire throughout the day. Of course, no star is then visible in the sky . **The whole earth has always been referred to as being scorched, but this is neither necessary nor correct. Corresponding with the parallel** passages dealt with below; nothing else is meant than the whole country of Ethiopia not the whole world.

According to Geffcken in various passages of text, "a world conflagration however, is of limited scope" and meant the whole country of Ethiopia alone (Ak. d. W. 1899). To make the text "readable" and its relationship with the legend of Phaethon's fall is a hopeless task because the text is nonsensical.

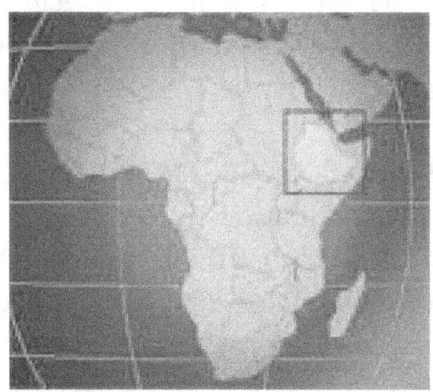

The modification of the tangled text would read: You inhabitants of India, do not think yourselves safe, and you too, proud Ethiopians! For when around this the flywheel of the ax of Capricorn and the bull with the twins rotates around the midpoint of the starry Southern heavens, when the Virgin ascends and the sun, with the girdle fastened around the forehead, then a great heavenly fire will arise on earth and many Ethiopians will perish in fire and groaning.

Now, at the time when the threatened event occurs, the axis of Capricorn is heading in a south-westerly direction, i.e. where, in Kugler's view, the Sibylline stars fight and Phaethon crashes from his place and plummets into Ethiopia. And in fact the maps of Herodotus and Eratosthenes (cf. Forbiger, Handbuch der Alten Geographie I, 68 and 180) testify that the geographical position of Ethiopia was assumed to be from Lower Egypt, ie. the home of poetry in a south-westerly direction. At the same time, Taurus and Gemini stand in the middle of heaven, in their common locations on the meridian.

The fact that the Virgin rises at the same time as the sun shows with full evidence that this is the same time of year as in the finale of Book V at the beginning of the battle, where Venus mounts the back of the lion as the morning star. **The girdle that the sun fastens around the forehead can be nothing but a dark band of horizon encircling the rising ball of the sun.** Our earlier interpretation of horizon is in harmony with this. In the light of the rising sun all the stars go out and Helios alone rules the whole sky.

Astronomical data of a certain day of the year on which a great upheaval (renewal) in nature occurs coincides with a fearsome heat that scorches the whole land of the Ethiopians. One should expect that the Indians are also mentioned. But their omission is certainly not because—as Geffcken thinks—"the bad poet forgets the Indians due to the often mentioned Ethiopians;" for one can be a weak poet without suffering from senile amnesia. If the whole passage were really nonsense and its author a feeble-minded dreamer, then one need not be surprised if after a few breaths he no longer recalled what he said. But just as the assumptions have proven to be quite wrong, the conclusion is also inadmissible.

Rather, the matter is as follows : "The whole country of the Ethiopians" united, as clearly emerges from Herodotus VII, 70, as two different races: the 'Ethiopians of sunrise' and those of Libya. The former, also called 'Ethiopians from Asia' by Herodotus, also differ by their language, the shape of their nose and their simply falling hair from the wool-headed Libyans, while both had the dark skin color in common (cf. Herodotus III, 101).

The 'Ethiopians of the rise' were assigned to the Indians of Xerxes' army. Understandably they were nothing more than a part of the

Kalatians, the dark-colored natives of India, who still live there, but at that time also spread much further west. The poet could therefore call the Ethiopians of the east 'Indians' with justification - in contrast to the actual (African) Ethiopians of the upper Nile region (cf. Herodotus III, 114). This is the only way one can understand how not only the Indians were "forgotten" when the poet speaks of the "whole country of the Ethiopians" in the various versions of Phaethon's fall. A collision which once set fire to the world and gave the Indians and Ethiopians the dark skin color noted in the Sibylline passage about the Phaethon saga in Nonnus, Dionysiaca XXXVIII. Geffcken's explanations certainly contain some valuable indications. But it seems that he saw great similarities where there was only a distant relationship, that on the other hand important real relations escaped him. The latter is of course related to his erroneous view of Sibyll,(V, 512 f).

3. WHO OR WHAT IS PHAETHON?

A pure fantasy? Difficult! In fact, repeated attempts have been made to uncover the natural process on which the legend is based (cf. Knaack in Röschers Lexikon III). Many believe it is a symbolization of the sunset. "Every evening the sun god falls in the west, and every evening the firmament and the mountains shine with red embers, as if the world were about to go up in flames. All that was needed was for this regularly recurring process to be understood as a unique event and for the sun god Helios-Phaethon to be treated or represented as a concrete reality, and the myth was complete."

Others, however, most recently Wilamowitz and Knaack, interpret Phaethon as the morning star. Ovid and Nonnos, however, agree with neither the one nor the other assumption. As far as the symbolization of the sunset is concerned, I cannot, of course, agree with Knaack's objection that the nickname of Helios is a "rather meaningless epithet" of Phaethon; it also seems to me that Phaethon II (son of Helios and Clymene the Titan goddess in Greek mythology, daughter of the Titans Oceanus and Tethys, thus making her an Oceanid) must be separated from Phaethon I (son of Eos and Cephalus who was a mortal prince in Greek mythology, famous for being the husband of Procris, the Athenian princess).

If I still reject the interpretation mentioned, it is because such an ordinary, simple and peaceful phenomenon as the sunset could not become the basis of a legend that obviously describes extraordinary, changeful and powerful natural processes. Likewise, the appearance of Venus as the morning star could not arouse even the boldest imagination of a world catastrophe. One might well imagine the morning star as the charioteer of Helios, or the evening star as a divinity thrown out of the sun chariot (cf. Knaack 1. c. Sp. 2178 f.). Likewise, the ascent of Venus's stem to its maximum elongation could be interpreted in a mythical way as a striving for kingship in the heavens. But a Phaethon in the sense of the 'Hesiodic' poetical or the Alexandrian version (which latter one has regarded as the basis of the representations of Ovid, Lukian, Nonnos, etc.; cf. Knaack 1. c. Sp. 2187) never came out of the Venusian planet.

But there is a natural phenomenon that could very easily become the reason for that legend. In the search for it, the following moments are to be considered as possible: 1. Phaethon appears not only as an epithet of Helios (sun god); he is also equated with Helios (especially in Nonnos). It is a myth of the sun or a sun-like celestial body made concrete. 2. Phaethon is not the driver of the sun chariot in which Helios is driving at the same time, but the former takes the latter's place. 3. The journey deviates in direction and speed from that of the sun. 4. The starry sky flares up brightly. 5. Phaethon is struck by lightning and falls down. 6. The flames of phaethon's conflagration also ignite the earth.

Certain meteor phenomena are now in complete agreement with this. Not only in modern times, but also in ancient times, meteors have been observed repeatedly, which were similar to the sun in terms of size and brilliance, crossed the sky in different directions at great speed, but then not infrequently exploded with thunder and lightning, sometimes also set fire to earthly dwelling places and fields with their glowing ruins. It is easy to understand that such an unexpected phenomenon, in the popular or poetic conception, should be confused with the Phaethon event. (Trans. note - the Tunguska explosion in the early 1900's is a prime comparison).

Oddly enough, the mythologists have not considered this very striking relationship between the saga and the appearance and effect of a meteoric sun, although Antiochenus (according to Plutarch) also attributed the saga to the descent of a ball of fire and also Valerius Flaccus,(Arg. V, 429 ff.) obviously thought of a meteor (globus ater) (cf. Knaack 1. c. Sp. 2193 f.). However, the relevant indications are not yet convincing; but they should have led to a more thorough examination, as suggested above. For us, however, the reason for this was by no means the two above references, but rather the nature of the matter itself.

At the same time, a surprising similarity between the Phaethon saga and the finale of Book V of the Sibyllines is revealed: here, as there, the star battle is introduced by the appearance of a brilliant meteor sun. There is also agreement insofar as the scene; in both cases it takes place in the morning sky. This is probably not so much because the meteors are visible on average much more frequently than usual at dawn, but to complete the Helios character of the Phaethon. In Dionysiaca XXXVHI, 365, as in Sibyl, there is a compensating reason: Venus as morning star (see next).

4. THE STAR STRUGGLE OF DIONYSIACA AND THAT OF SIBYLLINA. (Trans note - primarily astrological descriptions).

To justify the close relationship between the two poems by Dionysiaca and Sibyllina, Geffcken contents himself with referring to Dionys. XXXVIII, 356 ff., where the turmoil in the star world caused by Phaethon is mentioned. However, it should be noted that the whole character and the individual scenes of this 'battle' differ from those in Sibyll. V, 512 ff. and are quite different. This becomes clear from the following translation by Dionys: "And confusion reigned in the ether, and the order of the universe, which no one else shall stir, he (Phaethon) stirred up. Even the axis spinning through it was shaken by the swirling ether. Likewise, in Libya, the crooked Atlas, who supports the vaulted starry heavens, could with difficulty hold on to his squatting position; for the burden was too great for him. The dragon (Draco), which, with its twisted and curved belly, slid (his) circular path outside the great bear (Ursa Major - Big Dipper) in the same period of a day as the bull (Taurus), his star-rich companion, hissed at him, and the lion roared at the glowing gorges in the sky, heated the ether with mighty embers. The shaggy beast sprang up in a bold leap, and shooed the eight-legged crab (Cancer) away. On the star lion's posterior, its burning tail whipped the maiden, who drew near the path. But the winged maiden flitted past Arcturus, came close to the axis, and met the chariot. The morning star sent erring rays.

A literal translation is not possible at all in several places. Refraining from any translation, however, did not appear to be practical, since it encountered difficulties that even the trained Graecist will experience if he is not sufficiently familiar with the phenomena.

The meaning is this: As a star near the North Pole describes in the daily celestial movement an orbit which is much smaller than that of the bull, which is near the equator completing its day's journey with hasty steps, the clumsy dragon slides to its destination. Despite his awkwardness, the bull cannot escape him.

The dawn was on the run. Instead of the harmless hare grabbing the fiery Sirius after the greedy bear. The star-rich Pisces both left their place, one the south-west, the other the north-east, and hopped to Olympus, they, the neighbors of Aquarius. The ringed dolphin, Capricorn's companion, flipped over and began to dance. Before the scorpion (Scorpio), which was pushed aside from its course in the south in crooked lines and was coming near, Orion was still afraid at the height of the star because he grasped the scissors, since if he went only slowly he would have the second Draw the tips of the toes on the sharp sting (of the scorpion).

Luna also, standing at noon, rejected half the splendor of her face, turned dark, and sprang up. For she did not want to steal borrowed light from the light-bearer of the male sex and did not want to drink the splendor of her sisters from her counterpart Phaethon. From the crowd of Pleiades standing in a circle and screaming, a seven-fold, intertwined lament roared over the seven-belted.

The meaning can only be: in place of Venus, which shortly before (possibly on the previous day) had set as the evening star now appears as a morning star and also under the stars, ie not only on the (western) horizon, where it is about to be set by the scorpion that has risen in the east and is wounded with a poisonous sting. (Cf. the legend of Orion's death in Greek mythology).

As the scorpion leaves its usual place, it rises earlier; therefore Orion must hasten his downfall to escape the sting. The power of the moon (last quarter) catches on in the radiant splendor of the nearby Phaethon, i.e. the meteor that shines as bright as day. Loud noise rising from the same number of mouths, the stars ran in competition and were as if out of their minds in their wandering course. Jupiter pushed Venus, Mars, Kronos (Saturn), and my star (Mercury) came close to the spring equinox on its journey, and after it had been given to the seven stars by the related light, it rose half bright next to my mother Maya, turned away from the heavenly chariot, whose companion he always is, either as a forerunner in the morning, or in the evening, when the sun has set, sending his light from behind.

The astrologers gave it the name "Sun Power" because it follows the same course evenly along the path. Stretching its neck damp with dewy flakes, the bull of heaven, Europe's fiancé, roared, and at the same time raised its crooked foot to run. First he pressed the pointed horn to the side and then he dealt a blow to the wreath of the heavenly chariot with his burning "claws". Now Orion grew bold, and drew the sword from its scabbard by the burning shank. Arcturus ((red giant star known as the herdsman in the constellation Bootes)waved the crooked shepherd's crook. Pegasus whinnied, throwing up the star-barrel's knees in the air, and pounding the heavens with his hoof, the half-light Libyan horse ran to his neighbor the swan, flapping his wings with a snort, only to knock down another charioteer from the ether just as he had overthrown Bellerophontes (the hero in Greek mythology who overthrew the monster, Chimera) from the heavenly chariot.

No longer did the circling stars dance in the bear, hips clasped high up near the north pole, but turned south-west and bathed in the lake. Note that the whole description of the poet is put into the mouth of Mercury, who narrates it to Bacchus.

If one now compares this with the passage from the Sibylline explained above a profound difference is revealed both in the plan as a whole and in all the details of the battle. Nonnos has purposely designed a most fantastic picture of the utter confusion in the starry world; the author of Sibyl on the other hand, has shown real astral transformations, as they normally take place after a period of seven months, as the effect of a struggle. That's a very important difference. Gefficken seems to want to derive a resemblance from this, if we correctly understand his emphasis on some of the names of the stars that can be found both in Nonnos and in the passage of the Sibylline.

But that is not the point; for it goes without saying that some of the many stars that Nonnos mentions can also appear in any other astral poetry. The only thing that can be considered is the relationship between stars of the same name; in this respect, however, there is the greatest imaginable contrast between the two poems, even before the knowledge of the unified plan of Sibyl.

5. PHAETHON'S FALL INTO ERIDANUS (CELESTIAL RIVER) WORLD FIRE AND DEUCALION (GREEK NOAH) FLOOD.

As the confusion among the stars reaches its climax, Zeus (Jupiter) intervenes; he hurls his lightning bolts at the troublemaker and forces him down into (earthly) Eridanus. But then, like the latter, he is transferred to heaven as Auriga and the order of the world is restored. Apparently our Sibylline passage offers no parallel to this. And yet there is one. The sibyllist has - as already shown in part I above - known and used the Phaethon saga: his 'shining sun', whose threat he sees in the starry sky, is like Phaethon, a sun-like meteor. But the sibyllist also wants to not completely renounce the effective lightning scene of the legend. But Zeus - Jupiter - because pagan - must fall.

A natural symbol of divine judgment takes its place; the seer sees not only the threat of the "shining sun" but also "the terrible anger of a moon in lightning bolts." It is a moon-like meteor, which explodes under lightning and thunder. Now it becomes clear to us why two meteors appear in the prelude to the Sibylline star battle, of which only the meteor sun is brought into causal connection with the following star battle. While it is only said of her (but not of the 'moon') that 'long flames' (= fiery meteor tails) fight for her (or instead of her) expressly mentioned; but the upheaval in the starry world initiated by their appearance ends with the stars being hurled into the ocean. The lot of Phaethon's is thus transferred to the stars. Consequently, they also had to be presented as the direct cause of the earthly conflagration, while causing it indirectly. The Sibyllist describes the scene as 'the whole country of the Ethiopians'. The Dionysiaca XXXVIII44 says nothing about this.

And Ovid mentions the effect of the fire on the Ethiopians only in passing (Metam. II, 235): Sanguine tunc credunt in corpora summa vocato Aethiopum populous nigrum traxisse colorem (They then believe that the blood on the bodies of the Ethiopian people was changed to the black color). On the other hand, the older commentators on the passage in Plato, Timaeus 22, where the Phaethonean fire is mentioned alongside the Deucalion's flood, assume Ethiopia as the location of the former and Thessaly as the scene of the latter.

But one doesn't just let a saga play out there, but rather counts on actual events, if only in the sense that the saga started from them. What is most remarkable, however, is that several testimonies express this simultaneity of fire and flood. First of all, Celsus regards them as historical facts (in Origen, contra Celsum I, 19 and IV, 11) and Origen himself does not deny their historical character (cf. also IV, 21). On the other hand, it cannot be seen from the relevant passages whether the flood and the fire occurred simultaneously after Celsus and where the latter took place.

The statement in Tatian, Oratio ad Graecos 60 (Otto p. 150) is more specific: "The time of Cecropis, the fire of Phaethon and the flood of Deucalion", occurred together, which can be found in Clemens Alexandrinus, Stromata I (Potter p. 380; Stählin p. 66). In the same book (Potter p, 401; Stählin p. 84f.). However, Clemens also offers a chronological fixation of the double catastrophe, with reference to Thrasyllus (1st century A.D. Egyptian astrologer). He and others date the dual catastrophes taking place not before the capture of Troy in 1184 B.C., that is 330 (340?) years after the flood in Thessaly.

Many devastations happened in different places, as Plato relates" (477 after Abraham.). Augustine alone knows about the flood, De civitate Dei lib. XVIII, cp.10 (Migne t. 41 Sp. 568) to be reported especially after Marcus Varro: His temporibus, ut Varro scribit, regnante Atheniensibus Granao, successor Cecropis, utautem nostri Eusebius et Hieronymus, adhuc eodem Cecrope permanente, diluvium fuit, quod appellatum est Deucalion , eo quod ipse regnabat in earum terrarumpartibus, ubi maxime factum est. Hoc autem diluvii nequaquam ad Aegyptum atque ad eins vicina pervenit. (In those times, as Varro writes, during the reign of the Athenians, Granaus, the successor of Cecropis (mythological African king), was still reigning, when a deluge, which was called Deucalion's, took place in those parts of the world, extending to Egypt and its neighboring states.

Slightly differently, Paulus Orosius (the student and friend of Augustine) Adverse. Paganos I, 9: Anno DCGCX ante urbem conditam Amphictyon Athenistertius a Cecropis regnavit, cuius temporibus aquaruminluvies maiorem partem populorum Thessaliae absumpsit... Tunc etiam in Aethiopia pestes plurimasdirosque morbos paene usque ad desolation em exaestuavisse Plato testis est. (Before the founding of the city, Amphictyon Athenistertius was ruled by Cecropis, in whose time the flood of waters consumed the greater part of the people of Thessaly... At that time, even in Ethiopia, Plato witnessed that plagues and very many diseases were scorched away almost to the point of desolation.

Something different again Cyrillus Alexandrinus, Contra Julianum lib. I (Migne tom.76 col. 517). According to him, the 1st year of Cecropis falls on the 35th year of Moses, and in the 67th year of the latter —as they say— the Deucalionian flood and the fire of Phaethon, son of Helios, took place. Even if it doesn't occur to us to attribute a secure chronological value to these dates and to recognize the older chronology based on them (cf. Petavius, De doctrina temporum lib. XIII), we still have no right to the relevant one denying traditions any historical core. **If, however, there really was a fire and a flood catastrophe at the same time, then we have to look for a common cause of the two phenomena** - if we don't want to play chance into play.

But it has already been shown that the destruction caused by Phaethon is based on a meteor phenomenon; so the great flood is also due to the same. But is this possible? Without a doubt!

However, in order to properly appreciate our explanation, it is necessary first to become acquainted with some scientific facts.

1. The large meteors (balls of fire) do not always appear individually; they also appear in flocks or streams of immense breadth (thousands and millions of kilometers). So it can happen that regions of countries that are far apart are subject to the optical or mechanical effects of those celestial bodies at the same time.

2. The meteorites that reach Earth show the greatest variety in terms of size and speed. The larger the mass, the less the original cosmic speed (up to more than 60 km per second) is reduced by air resistance. Therefore, while small pieces hardly hit much harder than hailstones, the largest drive deep into the earth with terrible force. The effects of such a giant meteor are best illustrated by the Coon Butte meteor crater in central Arizona. The apparently volcanic 'crater' is 1150 meters in diameter 47 and 125 meters deep at its center, while the crater wall rises 40-50 meters above the surrounding plain. Around this wall, six and a half kilometers wide, is a belt of ejected sandstones, some of which are 20-30 meters thick even if at a distance of a kilometer away the meteor falls.

According to the traditions of the natives, one may even assume that the event is not all too distant. And now to our case! **From what has been said it is possible that one and the same meteor shower passed over Africa (particularly Ethiopia) and over the Aegean Sea, producing great fires there and violent tidal waves here.** Suppose that in a Thessalian bay a meteor like the one in Arizona fell, the devastating effect must be far greater, especially as a result of the expansive power of the masses of water vapor developed, which must produce a violent torrent. This explains the Deucalionian flood and its simultaneity with the burning of Phaethon. But this results in further connections between history and legend. **Phaethon, the originator of the world conflagration, also causes an earthly flood of water, overwhelming the Eri-danus.**

If our statements above are correct, then we must look for the latter in the Aegean Sea. And strangely enough, Sibyll agrees with that, stating Lesbos will perish forever in Eridanus. That Sibylline calls the Aegean Sea or at least a part of it Eridanus is all the more striking because he does not otherwise use this name. However, the latter is not borrowed from the geography of the ancients, it is of astral-mythical origin. It is true that Hesiod, Theogonia 338, already mentions him as the son of Okeanos and Tethys alongside the Nile and Alpheios; but the statements of the ancients about its earthly position are so variable and contradictory that the geographer Strabo rightly calls it a river that can be found nowhere (Geogra. V). Thus Herodotus III, 115 — who refrains from making any specific assumption himself — only gives the opinion that Eridanus pours into the sea 'towards the north wind', while Pliny and others identify him with Padus (Po) and Aponius Rhod. Argonautica IV, 627 ff describes it as the sea into which the Rhodanus (the Rhone) pours.

Eridanus is everywhere where the local legend places the fall of Phaethon, such as (e.g. B. Aponius Argon. IV, 597 ff.) lets the saga play out in the Celtic region. This difference in the scenes of the legend can be explained all the more easily since its natural basis - conspicuous meteor phenomena - with similar devastating effects. The message indicated above, that at the time of the giants God let a ball of fire fall down into the land of the Celts, which was extinguished after the land was devastated by Eridanus, an event on which the Greek Phaethon saga is based. It actually contains a core of truth, which is the basis of the presentation in Apollonius 1st c. that the Phaethon saga concludes with the fact that Phaethon's Auriga and likewise Eridanus were transferred to the heavens as constellations. Up there, however, there was such a body of water long before the Greek saga was formed, and this idea did not spring from the Hellenic spirit.

However, Herodotus III, 115 claims the name Eridanus is of Greek origin, but there are good reasons, i.e. he speaks for his Babylonian descent. For there is a constellation in the Babylonian firmament that bears the name of the ancient holy city of Eridu, which lay at the "mouth of the rivers" (Euphrates and Tigris) and to which, among other things, the Vela and the southern part of Puppis belong which, however, extended even further to the west in older times and thus came into direct contact with the constellation Eridanus. All the more so since Eridanus, according to Eratosthenes, extended far south to the area of the Canobos. And if one still wanted to doubt it, the fact is that before the founding of the city of Athens this area was ruled by Cecropis, in whose time the flood of waters consumed the greater part of the people of Thessaly. . . At that time, even in Ethiopia, Plato witnessed that plagues and very many diseases were scorched almost to the point of desolation.

In addition to these local catastrophes, antiquity also knows a fire and a flood of universal extent, but not on the basis of historical or legendary traditions, but based on cosmological speculation, according to which the world is dying in a periodic alternation, sometimes by fire, sometimes by water, only to be fully restored every time. This is what was taught in particular, which, however, essentially follows Democritus (cf. Alleg. Hom. c. 25) with regard to the nature of the annihilation of the world. The whole is based on the mutability of the elements. Just as the world came into being from the original fire, so, conversely, earth and water completely turn into fire again. (More details in Zeller, The Philosophy of the Greeks, 3rd edition, III, 1 p. 153).

This purification process has been thought of in different ways. Cleanthes said it emanated from the sun, since it contained the world-governing power (Plutarch, De comim. not. 31, 10). At the end of his *consolatio ad Marciam*, Seneca sketches a particularly lively picture of apocalypse: "And when the time has come when the world destroys itself in order to renew itself, then those (earth, life in the seas) will wear themselves out through their own strength (viribus ista se suis caedent), stars will bump into stars (sidera sideribus incurrent), and while everything material is aflame, whatever is now shining according to plausible distribution will rise in a single fire (uno igne ardebit)."

Seneca's description is even more vivid of the flood (Nat.Quaest. III, 27 ff.), which becomes a tremendous sea and forces itself up over the highest mountains and whose growth is not limited (solutus legibus sine modo fertur) for the same reason: "The one and the other occurs when God thinks it good that better things begin. Old things end. Water and fire dominate the earthly; from this their origin, thereby their downfall." This quite general philosophical-theological ratio is not enough for him; he also turns to the natural sciences of the time. **Among the information received, he likes the explanation of the Babylonian priest Berosos, according to which the two great events are caused by the course of the stars.**

The earth conflagration occurs when all the stars, which are now moving in different orbits, meet in the same point of Cancer, while the flood occurs when they meet in Capricorn.

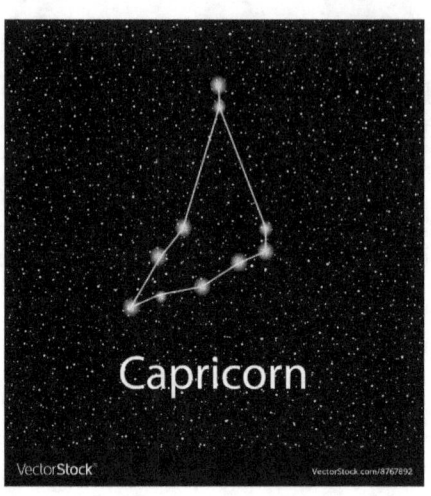

CAPRICORN

The solstitial points (first days of summer & winter) lie in the two constellations and this circumstance also gives great importance to the conjunctions mentioned. This agrees with what Censoring, (De die natali 18, 11,) says about the "great year": Annus, ... which the sun and moon, and the rambling planets, form the five stellar orbits, when they are related to the same sign, where they once existed together; the winter of this year is the most extreme catastrophe, which our people call a deluge. (quem solis et lunae vagarvunque quinque stella rmn orbes conficiunt,cum ad idem Signum, ubi quondam simu 1 fuerimt,una referuntur; cuius anni hiemps summa cataclysmos, quam nostri diluvionem vocant, aestas autem ecpyrosis, quod est mundi) or: which of the sun and moon and the rambling star form five orbs, when they are related to the same sign, where they once existed together; of which year the greatest winter cataclysmos, which our men call a deluge, and the summer of the ekpyrosis, which is the world).

According to Heraclitus and Censorin, it is not just the earth that is seized by the flood, but the whole cosmos; this also demands their consistently implemented system planets, not "all celestial bodies"' (as Geffcken, 1, cp 703) erroneously states. (52 dium). According to Nemesius of Emesa (IV./V. Century AD), this view was simply stoic doctrine". The same can be all the less surprising, As well as otherwise, there are intimate relationships between the views of the Stoics and their astrological contemporaries. Where does this come from? According to the fundamentals of astrology, all growth and decay in the world, be it large or small, is predetermined by the course of the stars, their forms of appearance and interrelationships.

The stars, especially the planets, are heralds of the divine will. As long as the unalterable regularity of the planetary course was not known, astrology could not acquire a deterministic character. But when, in the course of the last seven centuries BC, and prudent observations of the Babylonian priests, that after certain times the planets return to the same place, solar and lunar motions balance out, the lunar eclipses are repeated with regard to size, duration and course, that finally even the changing speeds of the wandering stars, in particular their maximum and minimum, are subject to temporally and spatially determined laws. In short after one had recognized that the movements of the heaven do not follow the free will decisions of merciful or unmerciful gods, but take place with absolute necessity, then astrology, like polytheism, either had to collapse or lead to a deterministic system that had to be regarded as pantheistic fatalism.

According to this - at least in principle - all future events in the world - be they physical or moral - could be determined with certainty from the present position of the stars by mathematical combinations. But such a determinism is peculiar to Stoa. Hence their fondness for astrology and their efforts in justifying it by pointing to the constant co-incidence of certain earthly and heavenly phenomena, that 'sympathy' of all parts of the world, from which the unity of the world and the primal force resulted.

Other similar catastrophes differ from the fire of Phaethon and the flood under Deucalion not only by their universality and the complete difference in the natural factors causing them, but also

by the fact that they do not take place simultaneously, but rather separated by huge periods of time and are repeated periodically. Likewise, the events in the starry world are of a completely different nature than in the case of the Phaethon fire. Here there is confusion, hostile pursuit and flight—like in a circus or menagerie fire, where every animal flees for self-preservation or—when this is not possible—blindly attacks weaker, more enraged neighbors. There, on the other hand, self-destruction, a collapse to accelerate the common death by fire determined by eternal fate; this is not a 'star fight'.

After all, it is by no means impossible that the two very different fires and floods were linked. C E Sus seems to have blamed the Christians for such an erroneous connection when he (Origen contra Celsum IV, 11(Koetschau p. 281) asserted: The alleged Error could only consist in the fact that they regarded the Deucalion Flood as the youngest of the great world tide signified by the return of the stars (planets) to the same point in the sky and, due to the lawful upheaval in the cosmos, expected a coming world fire. The accusation is of course unfounded (as Origen also shows). But it was closer to Celsus since within the younger Stoic school even the fire of Phaethon and the flood at the time of Deucalion were regarded as two of those great, periodically recurring catastrophes.

This emerges with full clarity in Manilius, the Stoic poet of the first imperial period, Astronomicon IV, 829 ff.: cc.: The earth is shaken, clinging to various joints, and he draws his feet on the ground. The world swims in him. And the ocean vomits up the sea, and thirsts after it sucks. No. He takes himself. Thus once the cities had plunged, when Deucalion became the sole heir of the human race, he possessed the whole world with a single rock. Nor did Phaethon attempt the reins of his native land, the nations burned, and the sky was afraid of the fire; Only in a long time do they change their course, and return again to themselves. Thus at a fixed time Symptoms also lose their strength and take Manilius as apparently at a loss to explain the two partial catastrophes; so he turns them into world catastrophes, but in contradiction to the stoic system that demands total annihilation he lets Deucalion live and lets the world languishing under the blazing heat of the Phaethonean fire get away with the terror of death.

Manilius is apparently at a loss to explain the two partial catastrophes; so he turns them into world catastrophes, but in contradiction to the stoic system that demands total annihilation he lets Deucalion live and lets the world languishing under the blazing heat of the Phaethontean fire get away with the terror of death. Seneca is wiser. Where he speaks of the great fire and flood catastrophes and their connection with the conjunction of the planets (Natur. Quest. III, 29 imd Gonsolatiad Marciam, 26) he mentions neither Phaethon nor Deukalion. And in other passages of his observations of nature, too, he does not think of them. This is all the more striking given that the Stoics went to great lengths to interpret a whole host of myths in their 'philosophical' sense. So already Zeno, Kleanthes and Chrysippus.

TRANSLATOR'S SYNOPSIS

Astronomically, Phaethon was a physical cosmic intruder into our solar system which - like in mythology - seemed to take the place of our sun because of its enormity and fiery makeup. It was accompanied on its rampage through our solar system by a smaller moon-like object which was surrounded by blazing, flaming tentacles and which was also attended by a mass of smaller cometary objects. The origin of these cosmic interlopers has yet to be discovered. But that they did exist and caused great destruction on Earth was recorded in various historical documents and poetical renditions.

There were 2 major cataclysms at the same time: the fiery crash of Phaethon with the Earth and a massive Deluge of water that accompanied it. The dating of the event is curious: in one source it is about 1100 B.C. in another it is 100 B.C.. This is difficult if not impossible to reconcile.

LIST OF PRIMARY FACTS

Judges 5:20 shows that the stars are seen as warriors. "Heaven's army" is therefore not just an analog, but means a real army of fighters. Where does this idea come from?

The orderly movement of the fixed stars and their ranks, which are revealed through the difference in shine and color, already suggest the representation of an army.

The apparently random special movement of the planets, which now move forward (to the east), now backward (to the west), now faster, now slower, now in a straight line, then in a tortuous path under the stars, and even stand still for a long time are like scrutinizing commanders.

"From heaven the stars fought, from their orbits they fought with Sisera." Sisera was commander of the Canaanite army of King Jabin of Hazor.

Particularly striking among the star warriors is the magnificent Orion, In Greek mythology, a gigantic, supernaturally powerful hunter who in any case awakens the idea of a mighty war hero.

These celestial sights evoked warlike ideas, especially when the planet (Jupiter) was in the middle of the ring's that circled the threatening moon, and the representative of a hostile power (Mars) approached him from outside: a picture of the siege and the threat of invasion.

These examples may suffice. They explain the view of the ancients that the starry sky is not only alive, but also a scene of warlike complications.

For us, however, we are not concerned with the appreciation of the poetic clothing, but with the actual, natural-historical basis of the two ancient written classics: Ovid's "Metamorphosis" and Book V of the Sibylline Oracles (Vs 512 ff.); and also the secret of Phaethon which was kept mostly hidden.

But once the first and most difficult riddle has been solved, we have also found the key to the second. Both will turn out to be fact in the course of our investigation. Ancient traditions, even in the guise of myth or legend, should not be easily dismissed as fantastic or even as meaningless creations.

1. THE SUPPOSED "MAD FINALE" OF THE 5TH BOOK OF THE SIBYLLINE ORACLES

The end of the fifth book (Vs 512-531) of the famous Sibylline oracles forms the prophecy of a "war of the stars" which ends with the fact that they will be hurled down and will set the whole earth on fire. So faithfully does the poet reproduce the astronomical processes that the modern tools of calculation not only mark the homeland of the poetry's origin, but even determine the season to which it corresponds.

Two mighty meteors of apparent size and shape of the sun and the moon (as they are seen at a distance) appear threateningly in the sky with their characteristic accompanying phenomena in turmoil; and the actual star battle begins.

The morning star (Venus), standing on the back of the lion (astrological Leo Constellation), initiates it. (Trans - this suggests that some unusual cosmic event caused Venus to act in an unorthodox way).

The stars that ruled the dawning sky at the beginning of the battle finally descend into the Oceanus (the river flowing around the world), setting the earth on fire. (Trans - the implication here is that some mighty celestial event caused a disruption of the normal functioning of the local solar system).

In undertaking this prosaic work, we cannot, above all, do without the astronomical calculation. It alone can give us certainty about the astronomical positioning. What is certain is that Venus stood as the morning star on the "lion's back" when the battle began; the sun was therefore east of this in Virgo, which is also attested by a parallel written source. (Trans - meaning that the sun was not in its ordinarily proper position due to whatever had affected Venus).

at the end of the battle, the interrelationship between the sun and the fixed stars, would give a time period of two thousand years ago.

2. EXPLANATION OF THE TEXT: THE PRELUDE TO THE STAR WAR

(Trans. - Kugler's chapter title above explains the text only in the sense that it fully and accurately describes the astrological concepts involved. It places the cosmic participants in their respective places amid the constellations that comprise the zodiacal panorama at the time of the war of the stars. I gleaned from this for review, significant pieces of relevant information, supplying specific dates, times and locations to the major events with which this saga is involved).

A shining 'sun' menaced the stars and I saw a 'moon' of terrible wrath flashing with lightning bolts. The stars were weaponized. God made them fight. Instead of the normal 'sun' there appeared only long flames tangled up.

The actual moon is a peaceful wanderer which does not cast lightning or roar with thunder. The meteor "moon" is different; it can do both. Sometimes one only sees flashes of lightning, but usually one also hears the mighty thunder of the bursting heavenly explosions. Understandably, this also causes the stars to tremble and extends the destruction.

The moon is the regular companion of the earth and has nothing to do with the moon-like meteor of destruction.

"The morning star directed the battle by mounting the lion's back." (The Lion is the constellation of Leo: its back is the eastern portion of the constellation). (Trans - what this means is that Venus, the morning star, was misplaced from its normal orbit which caused various other cosmic repercussions of negative nature. This further implies that Phaethon and its entourage of destruction probably crossed the orbital path of Venus in its path toward the Earth, meaning that they must have come from the direction of the inner solar system.)

Why is the morning star the leader of the battle and why does he mount the lion as such? Here we encounter ideas whose original homeland is to be sought in Babylon.

As already pointed out, the beginning of the struggle coincided with the time when the Sun was casually in the 15th degree of Virgo and the end when it had reached the 15th degree of Aries. **These two dates were (100 BC) 209.4 days or 7 synodic months and 2.7 days apart.**

The whole turning point in the stellar world begins and ends with the characteristic phases of the moon, the natural determinants of time, and the waning crescent moon is on top of and close to the morning star, which begins the battle.

Note the culprit; it is the ibex (the constellation of the horned goat). It is already over the meridian and fled west. But he does not escape vengeance. (Trans - the horned goat is known as a sacrificial entity).

"The Pleiades no longer shone." **This statement is very valuable as it gives another means of pinpointing the point at which the battle took place. Now the setting of the Pleiades in geographical latitude 30 took place on April 7th, 100 BC. on April 8 Julian, at a solar longitude of 13 (or 15) degrees.** Apparently the fight was over soon after.

We will therefore have to assume that the last day was around April 8th, or the day after the setting of r Tauri in the Pleiades at the same time as sunset.

The site of this location is therefore in Egypt (Cairo at latitude 30° and Alexandria about 31°).

The real sense - and this is what matters above all - can only be found in astronomy. We ask: what temporal-spatial relationship exists between Scorpio and Leo and does this fit into the already recognized plan of the myth?

The appearance of the head stars d, ß, n occurs in (101 BC) and the geographical latitude 30°

The heliacal sinking of Sirius took place at locations of latitude 30° in 100 BC on May 11; the end of the star battle, however, fell - as shown on p. 22 - on "May 8 or April July 27. (Trans - Deleting Sirius from the equation).

The prophecy of the star war with the following fiery catastrophe refers to Ethiopia.

The whole earth has always been referred to as being scorched, but this is neither necessary nor correct. Corresponding with the parallel passages dealt with below; nothing else is meant than the whole country of Ethiopia not the whole world.

According to Geffcken in various passages of text, "a world conflagration however, is of limited scope" and meant the whole country of Ethiopia alone (Ak. d. W. 1899). To make the text "readable" and its relationship with the legend of Phaethon's fall is a hopeless task because the text is nonsensical.

Now, at the time when the threatened event occurs, the axis of Capricorn is heading in a south-westerly direction, i.e. where, in Kugler's view, the Sibylline stars fight and Phaethon. crashes from his place and plummets into Ethiopia.

Astronomical data of a certain day of the year on which a great upheaval (renewal) in nature occurs coincides with a fearsome heat that scorches the whole land of the Ethiopians.

3. WHO OR WHAT IS PHAETHON?

Many interpret Phaethon as the morning star. Phaethon is not the driver of the sun chariot which Helios is driving at the same time, but the former (Phaethon) takes the latter's (Helios) place. Phaethon is struck by lightning and falls down and the flames of Phaethon's conflagration also ignite the earth.

Phaethon appears not only as an epithet of Helios (sun god); he is also equated with Helios (especially in Nonnos). It is a myth of the sun or a sun-like celestial body made concrete.

There is agreement insofar as the time of the heavenly attack; in both cases it takes place in the morning sky.

4. THE STAR STRUGGLE OF DIONYSIACA AND THAT OF SIBYLLINA

Trans - This covers only astrological descriptions and their meanings in those terms.

5. PHAETHON'S FALL INTO ERIDANUS (CELESTIAL RIVER) WORLD FIRE AND DEUCALION (GREEK NOAH) FLOOD

Two meteors appear in the prelude to the Sibylline star battle, of which only the meteor sun is brought into causal connection with the following star battle.

The upheaval in the starry world initiated by their appearance ends with the stars being hurled into the ocean. The lot of Phaethon's is thus transferred to the stars.

Where the Phaethontean fire is mentioned alongside the Deucalion's flood it is assumed Ethiopia is the location of the former and Thessaly (Greece) as the scene of the latter after the protecting mountains around it had been blasted away, allowing the deluge in.

Several testimonies express the simultaneity of fire and flood. First of all, Celsus regards them as historical facts (in Origen, contra Celsum I, 19 and IV, 11) and Origen himself does not deny their historical character (cf. also IV, 21).

The statement in Tatian, Oratio ad Graecos 60 (Otto p. 150) is more specific: "The time of Crocopris, the fire of Phaethon and the flood of Deucalion", occurred together, which can be found in Clemens Alexandrinus, Stromata I (Potter p. 380; Stählin p. 66). In the same book (Potter p, 401; Stählin p. 84f.).

Others date the dual catastrophes taking place not before the capture of Troy in 1184 B.C., that is 330 (340?) years after the flood in Thessaly.

But many devastations happened in different places, as Plato relates" (477 after Abraham).

If, however, there really was a fire and a flood catastrophe at the same time, then we have to look for a common cause of the two phenomena.

But it has already been shown that the destruction caused by Phaethon is based on a meteor phenomenon; so the great flood is also due to the same. But is this possible? Without a doubt!

It can happen that regions of countries that are far apart are subject to being struck by swarms of meteors with the effects of those celestial bodies felt at the same time.

From what has been said it is possible that one and the same meteor shower passed over Africa (particularly Ethiopia) and over the Aegean Sea, producing great fires there and violent tidal waves here.

Phaethon, the originator of the world conflagration, also causes an earthly flood of water, overwhelming the Eri-danus.

Sibyll agrees with that, stating that Lesbos will perish forever in Eridanus.

Eridanus is everywhere where the local legend places the fall of Phaethon, such as (e.g. B. Aponius Argon. IV, 597 ff.) lets the saga play out in the Celtic region.

The message indicated above, that at the time of the giant's God let a ball of fire fall down into the land of the Celts, which was extinguished after the land was devastated by Eridanus, an event on which the Greek Phaethon saga is based.

In addition to these local catastrophes, antiquity also knows a fire and a flood of universal extent, but not on the basis of historical or legendary traditions, but based on cosmological speculation, according to which the world is dying in a periodic alternation, sometimes by fire, sometimes by water, only to be fully restored every time.

According to this - at least in principle - all future events in the world - be they physical or moral - could be determined with certainty from the present position of the stars by mathematical combinations.

The earth conflagration occurs when all the stars, which are now moving in different orbits, meet in the same point of Cancer, while the flood occurs when they meet in Capricorn.

According to Heraclitus and Censorin, it is not just the earth that is seized by the flood, but the whole cosmos.

CHAPTER 1

www.ingramcontent.com/pod-product-compliance
Lightning Source LLC
Chambersburg PA
CBHW072111290426
44110CB00014B/1887